The Future of the Impossible:
the Physics and Ethics of Time Travel

by
Dannelle Shugart

Illustrated by
Zanna Heidrich and Joanne McAlpine

Dedication

I would like to dedicate this book to my family,
Michael and Ada.

To my mom, dad, brother and sister.

You are why my world works.

This page intentionally left blank.

Acknowledgement

A big shout-out to Zanna Heidrich who drew the original illustration in my master's thesis and to Joanne McAlpine for making the dream happen.

Many thanks to my Georgetown thesis advisor, Thomas McManus – an adjunct Profession and a NASA engineer.

To Georgetown and especially the SFS program for being the place where one can write a thesis on the physics and ethics of time travel.

Introduction

Why and How the Physics and Ethics of Time Travel are Interrelated

It is difficult to say today what will remain impossible tomorrow. The time has come to put our prejudices behind us and embrace the possibilities that lie ahead. The scientific community is slowly embracing time travel, as what used to be exclusively within the arena of science fiction becomes a force within the scientific community.

According to Theoretical Physics Professor Kip Thorne of California Technical University, it is reasonable to assume that an arbitrarily advanced civilization could have the capacity to travel through time.

His ideas on time travel have evolved from his work on wormhole travel, which theoretically allows both backward and forward movement along a timeline.

Wormhole travel is the only form of time travel that allows for movement both ways along the timeline. There are, however, other methods of time travel that proffer one-way journeys along a timeline. Here we will examine each of the methods of time travel, starting with those allowed by relativity theory and following with those anticipated by quantum theory. We will then examine the ethical implications of each.

To examine the ethical implications of the practical applications of time travel, it is first essential to understand the ways it may be possible to travel along a timeline. To do this, it is vital to understand both classical and quantum physics*.   Classical physics examines ways to time travel based on the special and general theories of relativity. Quantum physics, as it relates to time travel, revolves around black holes, wormholes, and a substance called quantum foam.

Albert Einstein's relativity theories indicate that time is not constant, but relative. This thesis will begin by viewing time through the lens of relativity. From this perspective, one can see the glimmer of time

---

* According to some, any theory that does not use quantum mechanics can be considered classical.

travel possibilities. Key theories include warp-drive, hyperdrive, Jupiter travel, the use of velocity, cosmic strings, and the implications of Gödel's universe, or a cylindrical universe.

Quantum theory postulates time travel theories. An important hypothesis described by quantum theory is travel via black holes. Other theories examined are wormholes, dimensional travel, and travel to alternate universes.

Chaos theory will be the binding agent between the physics and the ethics of time travel. The Butterfly Effect is the perfect agent for this, as it is both a scientifically proven theory and an ethically applicable method of determining the effects of actions. It is a key factor in understanding the physical effects of time travel on both the timeline and the universe.

This book will show that most forms of time travel are ethically sound. Time travel is no more harmful than any other means of travel. Once this level of safety is determined, this book evaluates individual cases and paradoxes. Evaluative factors measure according to their individual propensity to allow a traveler to invoke harm or inhibit the free will of another.

Contents

Chapter 1: Relativity Special and the General Theories of Relativity

As They Relate to Time Travel

Relativity theory is the backbone of many time travel scenarios. There are two theories of relativity, called the special and the general theories of relativity. Albert Einstein published his special theory in 1905 and the general theory in 1916.

Einstein's first theory is the special theory of relativity. It governs the physical relativity of all uniform motion (Einstein n.d., 59). This states that, assuming k is a Galilean reference body, "If relative to k, k 1 is a uniformly moving co-ordinate system devoid of rotation, then natural phenomena run their course with respect to k 1 according to exactly the same general laws as with respect to k"(Einstein n.d., 18).

Einstein used this theory as an expression of the natural law of the physical world. As long as there are constant velocity and direction of a mass while at the same time being devoid of rotation, the special theory of relativity explains how a traveling object moving uniformly and relatively to another object will obey the same laws of physics with respect to any other mass relative to it. The special theory of relativity is only assumed to be valid for objects in a state of uniform rectilinear and non-rotary motion with respect to k"(Einstein n.d., 61) where k is a Galilean reference-body.

In lay terms, the special theory works only under certain specific conditions. These special circumstances are such that at least two bodies with mass, called reference bodies, must exist. Both bodies must be moving at a constant velocity relative to one another; that is, they must be able to be examined in reference to each other. Additionally, neither reference body should be rotating or be within a rotating system. When the reference bodies meet these conditions, the special theory of relativity states that the laws of physics apply equally to both bodies. This simple concept has significant implications, although it has prohibitively small practical time travel applications due to its limited scope.

The special theory of relativity is limited in that it only deals with objects moving uniformly and with reference to one another, which becomes prohibitive when attempting to apply it to theoretical methods of time travel because it offers less real-world applications than a broader theory would offer. Even so, the special theory is an important matrix in understanding the physical possibilities of time travel. While most forms of time travel follow laws under the umbrella of the general theory, the special theory also allows theoretical forms, which is significant because it shows how viable time travel must be if it can be possible under such restrictive circumstances.

The general theory of relativity is an extension of the special theory; however, the general theory takes a grander approach to physical reality and begins with the following concept: "All bodies of reference k, k1, etc., are equivalent for the description of natural phenomena (formulation of the general laws of nature), whatever may be their state of motion" (Einstein n.d., 61). In other words, the general theory of relativity provides that any object traveling with any type of motion is bound to the same physical laws as any other object. The understanding of the velocity and mass of an object is relative to the objects around it. This is important to the study of time travel in that it follows that time is also relative to the objects around it or more reasonably, time is relative to the observer. This book will later develop this as an important concept in time travel.

Although the above are the most basic aspects of relativity theory, there are many other concepts integrated into this theory. The key features of relativity theory for those interested in time travel are concepts of time, the behavior of clocks, rotation, speed, and gravity. These five concepts discussed in this book are the backbone of many of the theoretical time travel possibilities.

According to relativity theory, time is relative to each individual observer. Einstein provides us with a thought experiment to illustrate this by utilizing two identical clocks (plus the implied synchronized clock held by an observer). In this experiment, one of these clocks is placed in the center of a circular disk while the other is placed on the outside of the disk. The disk is then set in a circular motion. All of this is seen by one observer located outside of the disk, but not moving as relative to the clock in the center of the disk, and all of this is being done inside a non-rotating Galilean reference body. At this point, you will have a clock with no motion (the one in the center) and a clock that is in motion relative to the observer (the one on the outside of the disk).

The clock in the center, the non-rotating one, represents the time of the observer. This is because the observer is not moving or rotating in reference to the center clock. The clock in the center of the disk, if synchronized with that of the observer's clock, will maintain its current rate of telling time. The observer will not notice any difference between the time on the center clock and the time on the observer's clock.

The rotating clock will illustrate the relative temporal change caused by rotation. This clock will tick slower than its counterpart located in the middle of the rotating disk. It will tick slower only in reference to the observer and the other clock, but from its own frame of reference, it will continue to tick and measure time in the same way as before. It is this temporal difference that is significant in understanding certain modes of time travel.

Einstein was concerned with the effect of clocks in motion, as opposed to clocks on "… a rotating body of reference," (Einstein n.d., viii). Several experiments have been conducted using planes to test his theories of clocks in motion. In these experiments, two identical clocks are synchronized, then separated. One clock remains on the ground, while the other is whisked away on a plane. Using sensitive measurement devices, the outcome is always the same (and as predicted by Einstein). The clock on the plane passes time slower than the clock on the ground. The time lag is very small but measurable. The faster the clock travels, the larger the time lag.

Clocks and gravity are intertwined within the general theory of relativity. Time kept by clocks is affected by gravitational pull and therefore, according to Einstein, "... a physical definition of time which is made directly with the aid of clocks has by no means the same degree of plausibility as in the special theory of relativity" (Einstein n.d., 98). It warps the space-time continuum, which means that it has the power to slow time.

In the special theory of relativity, time, as measured by clocks, is reliable and consistent. However, the general theory of relativity is not only broader and more appropriate to the realities of the physical world, but also provides for the malleability of time by gravity (and speed). This is the start of an understanding of the flexibility of the space-time continuum with which this thesis is concerned.

Clocks and speed are related by Einstein's general relativity in the same way that clocks and gravity are connected. Speed and time are inexorably intertwined in both relativity theory and time travel. When Einstein determined that the speed of light is a constant, it led to the conclusion that time is not a constant. A direct result of these concepts is the understanding that speed can relatively manipulate time.

Einstein's thought experiments are designed to illustrate the relativity of time. The concept can be extrapolated to relate to time travel. If clocks count time on an individual basis, then a person can expect time to be individualized to the person. Once the acceptance of individualized time has been achieved, certain types of time travel can more easily be understood.

Time and gravity are inexorably entwined as time and clocks, though in a different way. For hundreds of years, scientists used Isaac Newton's theory of gravity. This theory is based on and therefore requires, time and space to be absolute. Einstein replaced Newton's theory of gravity with Einstein's own equation, the General Theory of Gravity.

From The General Theory of Gravity, the gravitational mass of a body is the same as its inertial mass. An object's inertial mass is its mass at rest. All of the above must be understood in reference to the knowledge that inertia, weight, and heaviness are all the same parts of a body (Einstein n.d.,65).

Another concept of note within the special theory of relativity is that the length of an object is shorter when in motion than when it is not in motion, relative to an object not in motion. An object will continue to decrease in length as it speeds up. It is from the mathematics proving this point that one encounters a limiting velocity which, according to Einstein, "… can neither be reached nor exceeded by any real body" (Einstein n.d., 36). This limiting velocity is light speed. Interestingly, as the speed of an object approaches light speed, and length decreases, mass of the object increases. This makes it more difficult to accelerate. At the speed of light an object will have no length and infinite mass and will appear to be stopped in time. Unfortunately for time travel, objects with mass cannot reach light speed. Light can travel at such velocities because it is comprised of photons, and photons have a rest mass of zero.

The equation proving the concept of a limiting velocity is Einstein's most famous equation:

$$E = mc^2$$

It is this equation, energy equals mass times light speed squared, that rules an object with mass will encounter a velocity that can force its mass to become infinite.

According to John MacVey, author of Time Travel: A Guide to Journeys in the Fourth Dimension, a grand notion "… of the special theory of relativity is that the velocity of light in free space is the same in any direction as well as being the same for each and every observer. It is also quite independent of the motion of the body emitting the light"(MacVey 1990, 50). Because the velocity of light is the same for all observers regardless of their place or speed, it must be that time and space are inseparable and time is not absolute.

The special theory of relativity requires space-time to be flat. Flat space follows Euclidean geometry* and because of this, the special theory of relativity does not allow for most forms of time travel. However, the general theory is much broader and can accommodate other types of geometry**.

The general theory of relativity is much more forgiving in its laws governing the physics affecting time travel than is the special theory. John Gribbon states, "The equations of the general theory of relativity, the best description of space-time we have, explicitly allow for the possibility of time travel"(Gribbon 1992, 165). Although the general theory "allows" for time travel in many ways, one way

---

* Euclidean geometry is the geometry of flat space-time where the inside angles of a triangle always add up to 180 degrees. This is the geometry that children learn in high school.

** Where the inside angles of a triangle add up to something other than 180.

it can already be found in nature is in the effects of gravity.

The general theory of relativity is more comprehensive because it allows for observers moving relative to each other to travel at different velocities. Therefore, it is more pertinent to everyday reality than the special theory. It also explains the effects of gravity on space-time. The effects of gravity on space-time are key to many time travel scenarios. Gravity is also the link between general relativity theory, black holes, wormholes, and quantum theory. Each of these will be examined in turn.

Gravity is explained by relativity theory as being the force exerted by all objects of mass. Large objects have larger gravitational fields. Gravity is a force much like electromagnetism but on an infinitely smaller scale. Its effects are best seen on the cosmic scale. For example, although an apple produces a gravitational field, this field is negligible compared to the earth that is thousands of times larger and has a comparably larger gravitational field. Ergo, the apple will always appear to fall to the ground and the ground will never appear to fall to the apple. However, the apple falls towards the ground in proportion to the amount of gravity the ground exerts on the apple.

Gravity is fundamentally the measurable effect of

the warping of the space-time continuum by objects of mass. It is also explainable as pure acceleration. A clear way to visualize this concept is to imagine a bowl wrapped in plastic wrap. If you place a pebble on top of the plastic wrap, it will warp the wrap so that it dips under the weight of the pebble. The dipping of the plastic wrap is an effect much like the way space-time warps around objects. The dip effect of space-time will be more noticeable around an object the mass of the sun than it is around an object the mass of a human being (See figure 1).

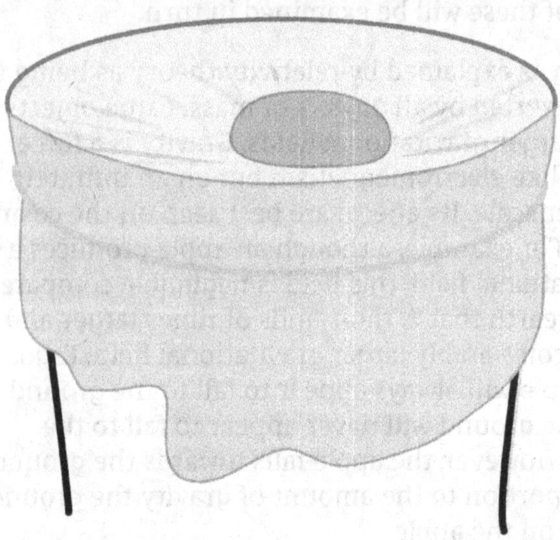

Figure 1: Effects of Gravity – demonstrated as a pebble on saran-wrap.

To understand why the gravitational warping of the space-time continuum is so important to time travel, it is important to understand exactly what the space-time continuum is. Space-time is the continuum of our three dimensions of space - length, depth, and width - attached to the fourth dimension of time. Of course, this thesis will later introduce ten, eleven, possibly even twenty-six total dimensions that should also be calculated into the space-time continuum.

Einstein firmly believed that "There is no such thing as an empty space …" (Einstein n.d., 155). To clarify, he went on to say, "Space-time does not claim existence on its own, but only as a structural quality of the field."(Einstein n.d., 155) As for space-time and its relationship to gravity: it is impossible to envision one without the other, literally. The example given above with the pebble on plastic wrap required a means of description, space-time, and a chance to describe gravity.

Einstein put his theories into mathematical equations. The importance of this is not to be underestimated. It is the curiosity and determination of the scientific community since the time these theories were published in the early 20th century that has begotten several sound theories of time travel. It is through trying to solve these equations using innovative and imaginative concepts that these novel ideas about time travel were born.

Scientists are not our only sources of scientific inspiration. Science fiction writers have provided other means of using the special effects of the geometry of the universe to scientific advantage in space and time. An example of this is warp drive. Warp drive is a concept that was popularized with Gene Rodenberry's Star Trek. However, members of the scientific community took the concept and used it as fuel for the academic fire of publications, experiments, and theories based on warp drive.

Warp drive is essentially the manipulation of space-time so that one can travel a geometrically altered path. That is to say, that warp drive is in effect a shortcut through both time and space. Miguel Alcubierre imagined the creation of a warp drive path using both positiveenergy-density material and exotic negative-energy-density material. By incorporating this material and using mathematics from the general theory of relativity, a tubular path could be warped in such a way as to create the proper geometry needed to produce a warp drive path. This path then could enable a properly equipped ship to traverse space-time in a non-linear fashion. A ship following this path could literally reach a region of space before a light beam, without bypassing light speed (Gott 2001, 124).

Another way to envision this concept is to imagine a long carpet runner in a cartoon. In this cartoon, a

Figure 2: Artist's rendition of warp drive

mouse is running away from a cat at light speed, but he won't outrun the cat because all the cat must do is shake the carpet in the right way and the mouse is drawn backwards as he runs forwards. It is the ripples in the cartoon carpet that emulate the manipulation of the space-time continuum when warp drive is used (see figure 2).

Hyperspace is a similar concept, but instead of moving space like a moving walkway, you go outside of space.

Hyperspace is the space outside our three-dimensional space. Imagine for a moment a two-dimensional world (often referred to as flatland). Anyone crossing from point A to point B in that world using three-dimensional space has an inherent advantage. The same is true for anyone living in a three-dimensional world crossing using hyperspace (See Figure 3). The advantage to a time traveler is time may not pass in hyperspace the same way as it does in real space.

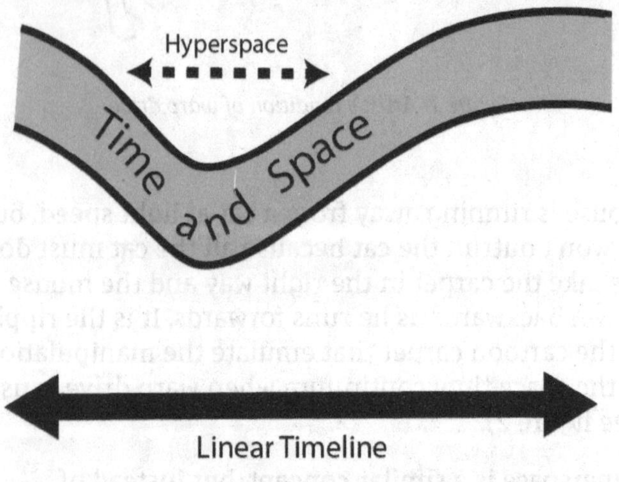

Figure 3: Graphical Representation of Hyperspace

Indeed, Dr. Parker suggests that "… scientists have speculated that time may not exist in hyperspace. Or, perhaps, it would not pass" (Parker n.d., 59). There is another theorized way to approach travel to the future without breaking the speed of light barrier.

It requires the ability to manipulate the mass of an object at least the size of Jupiter. Additionally, it would require knowledge of the size of the critical diameter needed for something of that particular mass to collapse into a black hole so that the time machine does not become a black hole.

The first step would be to create a spherical room for the time traveler to live in (referred to hereafter as the "Jupiter time machine"). This sphere would be larger than the critical diameter that would turn the living quarters into a black hole. The inside of this room would offer no gravitational effects to its occupants because the mass on the outside, being in a sphere, would counteract the gravitational effect.

The second step would be to take the rest of the mass that is being used and apply it evenly to the outside of the spherical room. Once a ball the mass of at least Jupiter was condensed tightly enough space-time would curve in such a way as to speed the traveler into the future (Gott 2001, 70).

The time travel involved in this situation is explainable using Einstein's theories. "… photons (particles of light) have energies that are inversely proportional to their wavelength: short-wave-length photons (as in X-rays) pack a large punch of energy, whereas long-wavelength photons (as in radio waves) carry just a little"(Gott 2001, 70). While safe inside the Jupiter time machine, the time traveler will continue to emit photons, as will anyone outside of the time machine. This is an important concept because photons are what light is made of. The photons emitted from within the time machine will need to expend energy to exit the gravitational well the time machine has created. As this photon (light) loses energy, it becomes redshifted.

Red shifting is when light's wavelengths become longer and the visual light turns red in the process. As a photon's wavelength elongates, an observer will see the light take longer to travel, because although photons travel at the speed of light regardless of the length of their waves, longer photons will take longer to pass any given point. Another way to look at it is that redshifted photons are like a yardstick and blue-shifted photons are like a ruler.

Even though both might travel at the same speed, the yardstick will always take longer to pass an object than a ruler will. Therefore, the person within the gravitational well of the Jupiter time machine will be seen by outside observers as aging more slowly because their emitted photons will take longer to be observed.

The time traveler will record the opposite effect. As the photons from people on the outside fall into the gravitational well of the time machine, they will invariably gain energy. These photons will blue-shift or turn blue and have shorter wavelengths. The time traveler then will see the people on the outside as moving more quickly. They will, in effect, age more quickly than our intrepid time traveler will. In order to speed up the temporal difference, the time traveler will only need to contract the mass. However, care must be taken to keep the shell of the room from collapsing; it will collapse if it is not at least four percent larger than it would need to be to become a black hole (Gott 2001, 74).

There is, possibly, an easier way to travel to the far future. Instead of manipulating gravity, it requires the manipulation of velocity. This manipulation of speed refers once again to Einstein's theory of relativity. Utilizing the philosophy that time is not a constant, but the speed of light is, it is possible to imagine that speed can take you to a future beyond time. Like the Jupiter time machine, a person can feasibly travel forwards in time in a way relative to an observer.

One issue that will need to be addressed is what is
referred to as the universal speed limit. According to
relativity theory, as an object approaches the speed
of light, its length shortens and its mass increases.
At the speed of light, the theory is that any object
that began with a rest mass of more than nothing
(zero) will reach infinite mass. Once infinite mass is
achieved, there is no way for an object to continue
to accelerate. At infinite mass, it would require
infinite energy to continue acceleration, and this is
impossible.

Cosmic strings, as opposed to the minuscule strings
in string theory, are endless and heavy. Like regular
strings , they may exist in either lines or loops.
Lines of cosmic strings are of infinite length (in an
infinite universe). Current theory holds that cosmic
strings "… should have a width narrower than an
atomic nucleus and a mass of about 10 million
billion tons per centimeter"(Gott 2001, 93). These
strings are stretched with tension. This causes them
to straighten with time and then lash out at speeds
often over half the speed of light. These cosmic
strings are so immense they would necessarily warp
space time.

To relate this to time travel, Dr. Richard J. Gott
proposes, in his book Time Travel in Einstein's
Universe The Physical Possibilities of Travel Through
Time, placing two cosmic string lines parallel to each

other where they will dramatically warp the space-time continuum. Using this warping, a time traveler could manipulate the gravitational effects if enough speed could be generated to move from one string to the other in a way using non Euclidean geometry to travel a shorter distance than the one which light will follow. The logical progression of this situation is that the traveler can arrive at the destination before the beam of light because a shorter distance was traversed, even though the velocity of light was never passed (Gott 2001, 92-98).

Time is one's interpretation of the light (photons emitted/seen/interpreted) around them. Time speeds up and slows down relative to what one experiences. Therefore, one will only live a specified lifetime. Even if one were to sit in a Jupiter time machine and time around was diluted so that the time traveler outlived several generations, this does not mean that the traveler will have gained several generations' worth of extra time. The traveler will not, for example, have gained enough time to read ten times as many books as any one lifetime would normally permit. However, assuming the traveler had the means to watch the civilization, the time traveler could potentially see books being read at percentage times the observer's normal speed.

One of the most fascinating aspects of physics is that it does not yet have all the answers to the questions of the universe. Some of these questions are: what shape is the universe? Is it static? Expanding, contracting, or rotating? While these are interesting questions for anyone, they are particularly pertinent to someone interested in time travel. The reason for this is simple; depending on the answers to these questions, there are other, specific forms of time travel available to future generations.

Assume for a moment that the universe is neither expanding nor contracting. Instead, imagine that the universe is of static size and is rotating. In this universe, it will be understood that galaxies are not moving away from each other but instead are in a fixed position from one another. In this universe, a non-rotating observer would see galaxies of sufficient distance from them as traveling faster than light as they spin around the observer. This would not conflict with Einstein's special relativity because "… the relative velocity of galaxies as they cross paths with each other cannot exceed the velocity of light" (Gott 2001, 91), and the galaxies in this model would never cross paths but continue in their static paths. This universe has been dubbed Gödel's Universe.

Gödel's Universe manipulates Einstein's theory of gravity by creating a situation where light is affected by the gravitational pull of the rotation of the universe.

Here, in this universe, light would travel not in a straight line, but in a curved line. As a result of light's unique trajectory in this universe, it would be possible to travel backwards in time.

Traveling backward in time in this universe would be a matter of creating a rocket ship that could travel fast enough to outmaneuver light. The ship would have to travel a different trajectory, of course. An easy way to envision it would be to think of a half oblong circle with a straight line cut through it. Light would travel a route along the outside of the circle while the ship, traveling at almost light speed, would traverse space along a straight line. At the destination the person in the rocket ship would arrive before the light beam which left at the same time and, in the world of physics, this is traveling backward in time. However, this type of universe, in all likelihood, does not exist.

Another time travel solution exists that is the result of a solution to Einstein's equations and concurrently assumes a specific type of universe that, in probability, does not exist. Frank Tipler discovered this possible time travel solution. His version of the universe includes "… an infinitely tall cylinder rotating at nearly the speed of light on its surface" (Gott 2001, 117). A shuttle flying around this cylinder could go back in time.

If the universe is finite and cylindrical, then it is conceivable that light might make a round trip all the way around the universe and end up where it started. This would happen if, for example, a stream of light left earth and sped through the galaxy on its way to the outer reaches of the universe. Then it might just end up at the other side of the universe. At that point, it is possible that it would somehow find its way looped back to where and whence it started.

Chapter 2: Quantum Theory

Wormholes and Alternate Universe

Black holes are where classical physics meet quantum physics in terms of time travel. In the classical sense, black holes are interesting in their ability to dramatically warp space-time. In the quantum sense, black holes are relevant when crossing the event horizon and approaching a singularity, so that, quantum theory replaces relativity theory in describing the best way to model the local reality. (More explanation on this later in this chapter.) Black holes are also a natural introduction to wormholes, which are themselves a unique method of time travel.

Gravitational pull and outward gas pressure provide the delicate balance in a living star. A black hole is created from the dissolution of that balance during the death of a massive star. The transformation happens in less than a second. As the star collapses, it changes color, reddens, and, finally, turns black. What was once at least eight solar masses (a star at least eight times the size of our sun), collapses into a point of infinite density, a singularity (Parker 1991, 132). A singularity is one of the most unusual things in the universe.

A singularity is a point at the center of a black hole where there is infinite density and infinite curvature of the space-time continuum. More specifically, according to physicist J. Richard Gott, "… quantum effects might limit the singularity's density to about $5\mathrm{x}10^{93}$ grams per cubic centimeter, but still, its size would be smaller than an atomic nucleus. Surrounding this tiny singularity would be just curved, empty space. Enclosing the singularity would be a spherical event horizon" (Gott 2001, 111).

The event horizon's radius is equal to the black hole's gravitational radius, usually only a few miles across. An event horizon is a dividing line between the place where escape out of a black hole is possible and where it is impossible. After crossing the event horizon, nothing, including light, can escape. It is the point of no return where it requires speed greater than light speed to pull free of the gravity of the singularity. Escaping the singularity requires this great velocity because even no mass photons (light) cannot breach a singularity due to the heavy curvature of space-time in this area. Therefore, there is no escape.

At the event horizon, time, as measured by anyone outside of the event horizon, stops. Additionally, when approaching the event horizon, an effect termed spaghettification occurs. Spaghettification was coined because of the type of stretching one would encounter as one approaches a black hole's event horizon (see figure 4). An intrepid time traveler, upon approaching the horizon would note an uncomfortable pull on her body. Assuming that, upon approach, her feet were directly pointed towards the middle of the singularity, she would find that the gravity pulling her feet would be much stronger than the gravity pulling her head into the singularity. Her head, in effect, would be slightly counter-pulled by the centrifugal force (assuming you are in orbit around the horizon). At the same moment, they will feel their shoulders squeezed together (as each is pulled towards the singularity point). At this point, our bold traveler will likely feel like a piece of spaghetti.

It is in part, due to these extreme conditions and the lack of black holes near Earth that singularities are so

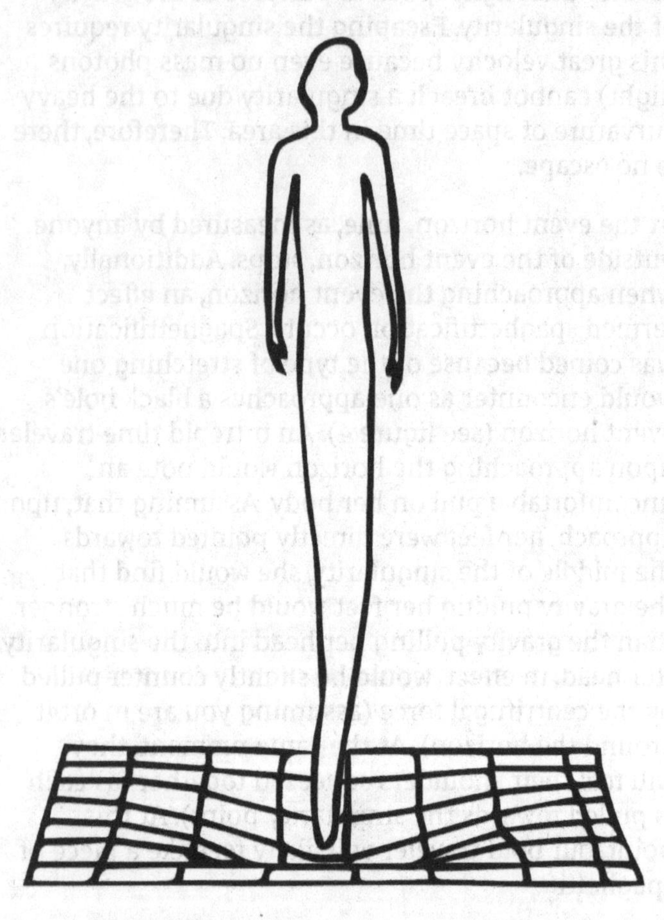

Figure 4 An artist's rendition of Spaghettification

hard to study. And this is precisely why there is not yet a theory of quantum gravity*.

There are five possible types of black holes: spinning, non-spinning, spinning with charge, spinning without charge, and primordial. Of these, only two are likely to appear in nature, Kerr (spinning) black holes and primordial black holes. However, Schwarzschild's black holes (non-spinning) are important in any theoretical discussion of time travel. Therefore, the following will be a look at Schwarzschild, Kerr, and primordial black holes as they relate to time travel.

Schwarzschild's black hole allows for a wormhole that would transport a traveler to another universe. According to Dr. Parker, "In this case we pass through two event horizons and then into another universe. In particular, we avoid the singularity"(Parker 2001, 139). In this way, an unusual type of time travel could be accomplished. While travel to another universe is not time travel in the conventional sense, it may still be considered time travel because any alternate universe is so unknowable there is potential for the said universe to be parallel to ours, yet at a different point of evolution.

A Kerr black hole is differentiated from a Schwarzschild not only in that it spins but also

---

*Quantum Gravity is the force of gravity at such a place where the laws of classical gravity no

because in addition to its event horizon it has a static limit. The static limit is a, according to Dr. Parker, "surface outside the event horizon that touched it at the spin axes and had maximum separation at the equator"(Parker 2001, 141). Primordial black holes are miniature black holes.

This type of black hole is particularly uncommon for two reasons. The first is that primordial black holes were only created during the big bang; none have been created since. Secondly, this type of black hole evaporates and shrinks, eventually exploding when a critically small mass is reached (Thorne 1994, 51).

There is a second type of time travel associated with black holes, due to time dilation from strong gravitational forces. Gravitational forces act much like velocity in relation to time travel. These are the forces that are used in the Jupiter time machine that was discussed in chapter one.

Although there are not yet known laws governing quantum gravity, many scientists are working on uncovering them. The laws of quantum gravity would describe many phenomena. They would be used to explain the probabilities (a la Schrodinger's cat) for the geometry of a singularity. The laws could also possibly anticipate or calculate the possibility that a singularity would create a new universe (Thorne 1994, 478).

Based on what we do know (or can reasonably interpret or interpolate), gravity begins to exhibit uncertainties near a singularity at such a scale that they become important. According to Dr. Gott, "At this density (5x1093 grams per cubic centimeter), quantum uncertainties in the geometry of space-time become important; space-time is no longer smooth, but instead becomes a complicated spongelike space-time foam"(Gott 2001, 162).

The foam, also referred to as Planck foam (see figure 5), found at the quantum level is the primordial mess where reality as we know it breaks down.   At

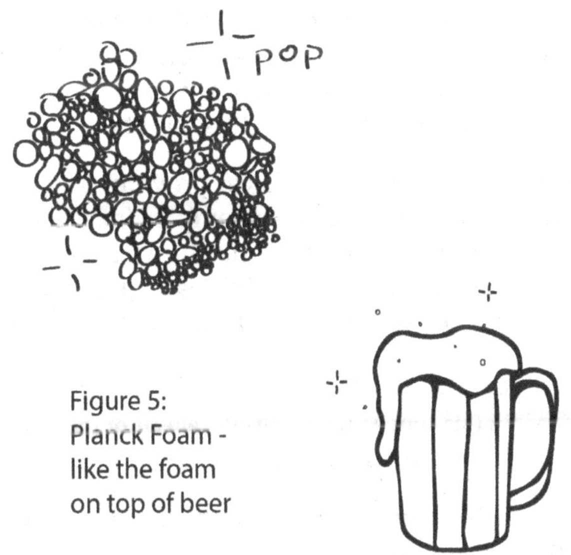

Figure 5:
Planck Foam -
like the foam
on top of beer

the quantum level, time is often measured in Planck time. This time is approximately a ten-million-trillion-trillion trillionth of a second, or 10^{-36} •

At this level, reality breaks down into uncertainty.

To explain this quantum uncertainty, consider the tale of Schrodinger's cat: as it goes, a scientific experiment places a cat and a capsule of cyanide in a box with no windows. Inside the closed box, the cat could be alive or dead; it is only when the experimenter opens the box, that the fate of the cat is determined. According to quantum theory, the cat is both alive and dead until the experimenter opens the lid. More accurately, there is only the probability that the cat is either alive or dead until such a time as the experimenter interacts with the experiment and forces a conclusion as to the cat's fatality.

As an analogy, Schrodinger's cat provides a reasonable visualization for a complex idea. Probability is at the heart of quantum theory. At this level, reality is not but fuzziness[7], quantum tunneling[8], and the probability that something may happen.

---

[7] Fuzziness in this context represents the effects of the Heisenburg Uncertainty Principle.

[8] Particles on the subatomic level appear to jump from one place to another without moving.

It is from this crazy mix that wormholes are created. A wormhole is a space-time portal through curved space. It has two mouths that can be manipulated to open not only in different places but at different times as well. Wormholes are postulated to be both microscopic and short-lived in nature, but theoretical physics provides ways of enlarging and maintaining them with the use of exotic matter.

Once a wormhole has been enlarged and stabilized, there is a way to turn it into a time machine. This requires anchoring one mouth while taking the other and accelerating it into the future (see figure 6). Attaching one end of the wormhole to a rocket ship and accelerating it into the future is one means of accomplishing this.

This version of a time machine provides time travel to the future. It also allows for time travel back. It does not, however, provide time travel to a time before the wormhole was created. This is sometimes given as the reason visitors from the future have never been encountered; a wormhole time machine has not yet been created.

Figure 6: An artist's rendition of a wormhole with one stationary mouth and one mouth moving at near light speed.

The differences between classical and quantum

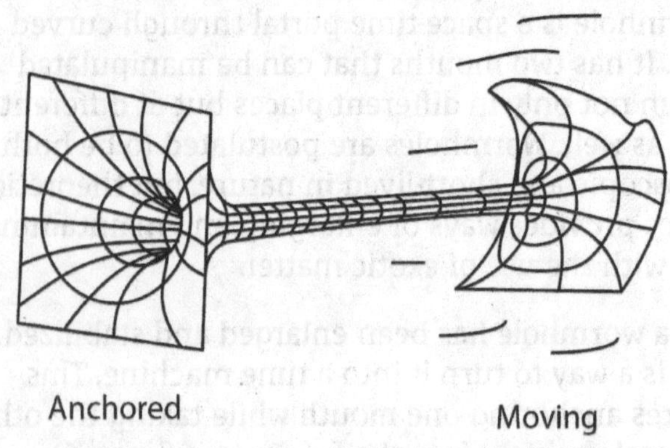

Figure 6: An artist's rendition of a wormhole with one stationary mouth and one mouth moving at near light speed.

physics are intense in magnitude and enormous in their implications of our understanding of reality. Classical physics deals with the physical world as we live in it. The effects of quantum physics begin at the most basic, smallest parts of reality (subatomic). At this level, the laws of classical physics, including Einstein's theories of relativity, break down and are no longer valid.

Scientists have been searching for a theory to encompass all our physical reality in just a few equations, combining classical and quantum physics. The name of this scientific Holy Grail is the Theory of

Everything. Physicists have noted that, as they learn
more about the way the universe functions, it seems
increasingly likely that they will eventually find
that the physical world is interrelated in such a way
that a theory like this will be possible. This Theory
of Everything, when complete, should explain the
laws of physics in such a way as to incorporate both
classical and quantum gravity.

A strong contender for inclusion in this Theory of
Everything is superstring theory. Superstring theory
is the unification of five different string theories. It
eventually became clear that each of these string
theories was a different way of describing the same
theory.

However, if string theory eventually incorporates
all five theories and becomes the best description
of reality, it may be that reality as we know it is a
fabrication. More accurately, it will be determined
that the space-time that we perceive is a mere
conglomeration of a more basic reality. Dr. Brian
Greene of Columbia University believes that because
each of the different sting theories requires a
different type of geometry, while concurrently all
describing the same reality. He believes that this
implies that our current geometry is a limiting factor
in understanding reality.

String theories are one of two kinds. One type

proposes that strings are closed loops while the other postulates that the strings are open. Each of these string theories, whether with or without closed loops, proposes that the basic building blocks of reality are composed of strings. These strings are energy, existing only in the dimensions of time and length. They do not have width or depth. Superstring theory anticipates the existence of ten, eleven, or twenty-six dimensions.

The first three of these are commonly used and understood as part of our reality's geometry. Some, however, do not understand why time is considered a dimension. One way to envision time as a dimension is to try to plot the location of your car. You might say that your car was parked at the top of your driveway, halfway between each of the sidewalls and that it was a foot from the door. This would only be valid during the time that the car was in the garage. Therefore, it is clear that time becomes a necessary plotting dimension.

The remaining dimensions are less common, less visible, and less understandable. These dimensions are at the forefront of scientific thought as we enter the 21st century. Although these dimensions are not part of common knowledge, they hold a good deal of promise for time travel.

These remaining dimensions would not be about

scale as the ones above (these remaining dimensions would be on the Planck scale), yet at least one of these dimensions would potentially be of use to a future time traveler. These extra dimensions are envisioned as loops of string. In fact, seven microscopic dimensions are predicted by superstring theory. Each of these dimensions are then predicted to be no larger than a loop with a circumference the size of 10^{-33} centimeter. According to Dr. Gott, "One of the extra dimensions could explain electrodynamics, a la Kaluza-Klein theory*, while the others could explain the weak and strong nuclear forces, which are responsible for some types of radioactive decay and for binding the atomic nucleus together. Just as every position along the vertical dimension of a straw is not a point but rather a tiny circle, every position in space in our universe would represent not a point but a tiny, complicated, seven-dimensional space 10^{-33} centimeters in circumference"(Gott 2001, 64).

To understand how a dimension can be so small as to be invisible to the eye and yet be in many places at once, consider a broken glass. As the glass lies shattered on the kitchen floor, the tiny glass shards can be imagined as another dimension. Each little

---

* The Kaluza-Klein theory explains light as a vibration of the fifth dimension which is what allows it to travel through a vacuum.

piece is a separate entity but is essentially part of the whole. Dimensions five through eleven can be imagined in much the same way.

Unified field theory was another attempt at a Theory of Everything. Unified field theory was a brainchild of Albert Einstein. It was meant to be a "… theory that would explain all the familiar forces found in nature, including light and gravity"(Kaku 1994, 98). It was unsuccessful.

A proposition of cosmic time travel supposes that the universe could use time travel to create itself. The advantage to this theory of creation is that it provides an answer to the age-old question "What was before?" Here, the before is part of a time loop.

This theory requires you to accept a version of the multiple universe theory. It incorporates a theory called chaotic inflation*. where it uses Quantum Theory to explain that as an inflating universe expands, it could produce baby universes. These universes in turn would expand and create more baby universes also sometimes known as pocket universes.

Li-Xin Li and J. Richard Gott proposed this theory in their paper "Can the Universe Create Itself?"

---

* Chaotic inflation is when quantum fluctuations cause spacetime to go to a higher vacuum energy density and a higher rate of inflation.

published in Physical Review Din May 1998 (and explained in Gott's book Time Travel in Einstein's Universe), hypothesizes that the start of this type of universe creation could be conceived within a time travel loop (Gott and Li 1998)

The idea is that during the first moments of creation, a time loop occurred, creating a situation in which the universe created itself (see figure 7). This time loop would be but a fraction of a nanosecond in time and a fraction of a foot in length (Gott 2001, 189).

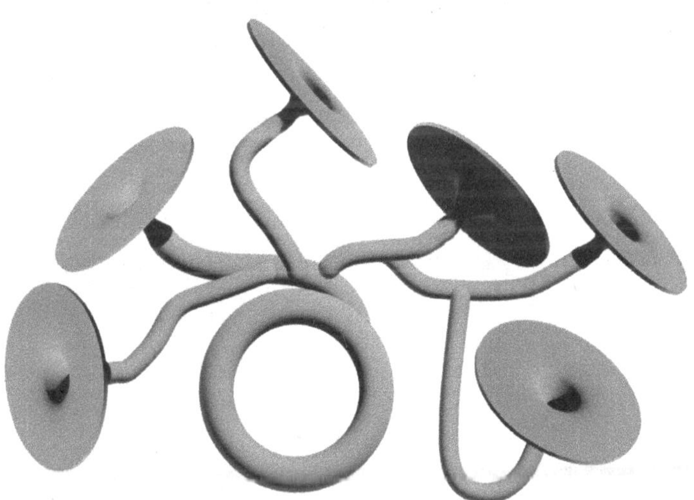

Figure 7: Self-creating universe

Besides baby universes being created by chaotic inflation, Lee Smolin of Pennsylvania State University has proposed that baby universes may form from black holes.

These universes might be created every time a black hole is formed; in this case, they would be separated from the parent universe by the black hole and its event horizon (Gott 2001, 192).

Garriga and Vilenkin have presented another baby universe theory: in their theory, baby universes are created from bubbles of high-density vacuum that form (in quantum foam) (see figure 8); assuming

Figure 8: Graphical Representation of a Bubble Universe

the parent universe" possesses a tiny cosmological constant"(Gott 2001, 192).

Each of these multiple universe theories can be incorporated into Gott's original time loop of cosmic creation and are in line with superstring theory. This is due to the fact that both superstring theory and the multiple universe theories presented here require that, in the beginning, "… all spatial dimensions were curled up and small" (Gott 2001, 192).

Quantum mechanics is unarguably the strangest discipline of physics. Consider the following concepts: eleven dimensions, results of scientific experiments altered by the mere fact that they are observed, particles that jump through walls. The above are phenomena that fall under quantum mechanics. Quantum mechanics is used to explain the behavior and nature of electrons, atoms, molecules, particles, and waves.

Heisenberg's uncertainty principle explains that we can never know a particle's position and its velocity concurrently is incorporated into quantum mechanics and has implications when considering time travel. This uncertainty principle has little effect on the macroscopic details of everyday life: however, at the Planck scale, it becomes significant.

The uncertainty principle deals specifically with

measurement: as you measure the position of an object, one will necessarily alter the movement of said object thereby destroying the accuracy of the original measurement.

According to Dr. Thorne, "The uncertainty principle is intimately related to wave/particle duality that is, to the propensity of particles to act sometimes like waves and sometimes like particles"(Thorne 1994, 373).

Quantum mechanics also contribute to the many-worlds theory as investigated earlier in this book. The many-worlds theory represents the possibility that reality branches off for every action taken: any action from choosing to brush your teeth in the morning to a physicist deciding to use a black pen to record her experimental findings will create alternate but parallel universes. The universes will represent each possible outcome from each decision, these universes will exist within different dimensions and never intersect.

As presented here, the many-worlds theory is important is because it provides the potential for another type of time travel, which is different from the rest because it involves what might almost be called a sideways jump.

If there are parallel universes, one for each action

taken, then perhaps it would be possible to puncture the barrier between them. If one could travel between universes, then one could potentially go to a place that represents one of the possible pasts of the current universe. The traveler would not be traveling backward in time as much as sideways in time. However, if the universe were chosen with great care, there would be nothing in this traveled universe that could make the traveler aware that she was anywhere besides her own past in her own universe.

Everything in the parallel universe would be identical to the specific past time chosen, with one very significant exception. In this universe, a traveler would come from another place and make an impact. An important concept to identify here is that the time traveler would probably impact whatever time or place she ended up in. From this observation, chaos theory requires that the ripple effects from the moment of arrival will substantively affect that time/universe from that point. Chaos theory will be discussed in greater detail in chapter three.

In the universe the traveler came from, the chaos effect would have no impact because all of the traveler's actions would be affecting a place in another dimension. Any choice or action in that dimension will have absolutely no effect on any other dimension, unless or until the traveler returns to the original universe.

The universe the traveler found herself in would feel the effects of chaos theory because that is the place she would be impacting. Although the chaos or ripple effect may present mammoth changes, there would be nothing inherently wrong with those changes, for although this universe may now contain two versions of the same courageous traveler, there is no catastrophic violation of the laws of physics.

Chapter 3: Chaos Theory
A Bridge Between Science and Ethics

Chaos theory is one of the most interdisciplinary theories that science has to offer. This theory weaves a web of stunning intricacies. Each pull on a string in this web provides a cascade of effects that start small but can develop a snowball-like effect that produces immense outcomes. Chaos theory is an important bridge between the science of time travel and the ethics involved in the implementation of time travel. A thorough understanding of the science of chaos will allow for a greater understanding of the implications of time travel.

Chaos theory provides insights on things from weather forecasting to time travel. Scientists studying population cycles utilize its calculations; even the basic swing of a pendulum is not out of its calculated web. Most importantly, any scientist truly interested in time travel will have to have a basic understanding of chaos theory.

There are two important yet significantly diverse patterns to chaos theory and the differences between them lie in the manner of the pattern development. Patterns can develop either with randomness that converges into a stable pattern or with randomness that increases with time.

The first type of chaos follows a pattern that almost appears contrary to its name. This pattern assumes a primary randomness at the start that then proceeds into patterns of chaos that converge to produce order. In Ian Stewart's words, "So chaos is a strange and beautiful combination of unpredictability and stability"(Stewart 2002, 130).

Chaotic systems often have underlying patterns of regularity. However, these patterns are often hidden and sometimes may appear to be random (Stewart 2002, 279). Therefore, it may be difficult to determine what type of system is being observed if there is not a clear understanding of chaos theory.

The latter is the one most often associated with chaos theory. It is a pattern of complexity that arises like the phoenix from two or more cycles of turbulence. They begin with apparent synchronicity, yet, due to minute differences in initial conditions, these cycles develop irreversible, wildly divergent paths. In time, the pattern changes so dramatically that the future of the path cannot be determined from its beginning path.

The power in chaos theory is such that, with movement, the predictive ability is lost with exponential rapidness. This means that in a system, such as the weather, an analyst has the power to make reasonably accurate short term predictions, but long-term predictions become increasingly difficult. The reason behind this problem stays the same whether it references the weather, the stock market, or issues in time travel. The reason is the sensitive dependence on initial conditions. This means that small changes at the start of a system grow exponentially with every change and modification until, finally, the system itself loses all semblance of its original form.

Equally influential to sensitive dependence on initial conditions is chaos' ability to create truly random motion. This can be seen in a spherical pendulum. Alternately, extremely unstable populations often settle down into more routine cycles and that is also a form of chaos.

An elegant description of chaos theory comes from James Gleick: "Tiny differences in input could quickly become overwhelming differences in output-a phenomenon given the name 'sensitive dependence on initial conditions'"(Gleick 1987, 8). This is a scientific way of stating that small changes or differences at the beginning, or initial condition of an event, can produce massive changes at the outcome or conclusion.

Pendulums are great toys for scientists, besides being wonderful illustrators of chaos. Vital in the first clocks and now assisting in revealing the patterns of chaos, they are a beautifully straightforward way to illustrate chaos out of simplicity and can provide unpredictability from regularity. Pendulums provided access to the idea that "… the disorderly behavior of simple systems acted as a creative process. It generated complexity: richly organized patterns, sometimes stable and sometimes unstable, sometimes finite and sometimes infinite"(Gleick 1987, 43).

There are several different types of pendulums. Yet for each type, according to Ian Percival: "The pendulum obeys Newton's deterministic laws, but any attempt to predict its future behavior over long times will be defeated"(Percival 1993, 13). Nonetheless, this is just an error in our ability to know and measure exactly not only the initial conditions but the effect of the driving motion and any other influences. It is an effect much like the Heisenberg uncertainty principle. Regardless, pendulums are still considered deterministic, even though they are chaotic, and this concept is termed deterministic chaos.

An effective way of visualizing deterministic chaos is with fractals that are shapes iterated upon themselves numerous times. Fractals are a way of proving what mathematicians are constantly and consistently saying; math is beautiful, or more specifically, chaos is beautiful. Benoit Mandelbrot coined the term fractal in 1975 from the Latin fractus, which refers to a broken and irregular rock. Many fractals themselves are self-similar shapes. Benoit Mandelbrot observed that "… they are irregular all over. Secondly, they have the same degree of irregularity on all scales"(Hall 1993, 123).

Fractals are an important natural shape and allow us access to understand the complexities of coastlines, for example, the chaos in fractals allows engineers to measure the complexities of nature. So how do fractals relate to time travel? The science is still in early stages, but already pioneers have stated, "… another area where fractals provide an apt description are in living things and in the Universe at large … counts of galaxies yield undisputed evidence that at comparatively small scales the distribution is fractal"(Hall 1993, 132). Fractals provide a beautiful visual of the Butterfly Effect (see figure 9).

Deterministic chaos may be best understood with insight into modern weather forecasting. In 1961, Edward Lorenz, a meteorologist and mathematician, was doing computer modeling involving weather patterns; one day he took a shortcut, placing in rounded-off initial conditions, and he expected it to have no effect on the outcome. His rationale for this belief was that the difference between the numbers he entered and the numbers that the computer would have entered would have had an insignificant .000127 difference. Such a difference should have been inconsequential; instead, he found that his weather patterns diverged until finally all similarity to the original model had vanished (Gleick 2001, 16). With this accidental discovery, the Butterfly Effect was born.

The Butterfly Effect, of course, has obvious implications to time travel. The popular description whereby a butterfly flapping its wings in China affects major weather patterns in Chicago is not as extreme as it first appears. While it is clear that the flap of small wings will not immediately alter major weather patterns halfway across the world, it is accurate to expect such variations will eventually have an impact. When transporting into another time or universe, it is imperative to understand the implications of even the smallest action.

Travels to the past invariably invoke fears of the impact of the Butterfly Effect, as they should.

Deterministic chaos implies that any entry into the past on a natural timeline (not into an alternate universe) will propagate unknowable changes to the future. Once the traveler has entered into their own past, they are compromising their original future. Power of this magnitude, affecting all life along a timeline, that the traveler should not enter into the past lightly.

Chapter 4: Ethics and Paradoxes

Physics shows that at a later point in our history, humankind may develop the ability to travel through time. This travel may take many different forms and each harbors a latent capacity for producing ethical dilemmas and so this chapter aims is to present the philosophical and ethical issues inherent in each form. Additionally, this chapter provides a framework for understanding which questions are pertinent in determining if a type of time travel is ethical.

To understand the implications of time travel, there

must first be a consensus on time itself. According to Dr. Brian Greene of Columbia University, space-time may not be the elemental feature of reality that has been assumed since science has been applied to the classification, measurement, and interpretation of time (Greene 2004, 477). It may instead be a macroscopic interpretation of a microscopic feature of reality. Or, as Julian Barbour argues in The End of Time: The Next Revolution in Physics, time does not exist at all. Alternately, time may be real, and exist, but may not flow in the manner that is generally assumed today.

The specific premise behind the idea that time does not flow is rooted in relativity theory. Imagine for a moment that there is a human on planet Earth and an alien on a planet a hundred million light-years away and that the alien has a specially constructed telescope. This telescope gives the alien the ability to see a hundred million light-years away. Now suppose that the alien is sitting and watching a human in a soccer game played on March 26, 2004. After watching the game for a few minutes, the alien walks a mile away and gets out her telescope to continue watching the game. However, relativity theory thwarts her plan because the alien is now watching Earth at a much later point in Earth's time. This is possible because when the alien took those steps, it put her significantly farther away from Earth (much like a small adjustment in the placement of a knife can make a significant difference in the size of a slice of cake) and this translated into a large difference in perspective between the alien and the soccer game on Earth (Greene 2004, 134- 136)

If we translate a small moment in time into a giant moment in time by a small movement in space, then how can time flow? If an alien can observe both the beginning and end of a person's life, then perhaps that life is suspended in time. Perhaps all of time is but a large chocolate cake of which any slice may be observed given the proper perspective.

If it is true that moments are suspended in time, can free will exist? If a distant voyeur can observe every moment in a life, how can choice be anything but an illusion? It must be that if choice is an illusion there can be no free will, for free will is the freedom of mind to choose that which is morally correct. If there is no free will, then the possibility of morality is moot. For if one cannot choose to be good and one cannot choose to be evil, then morality is but an illusion.

Morality is one of the cornerstone concepts for Immanuel Kant's belief that God exists. This rests on the theory that if God is omnipotent, yet evil exists in the world, then there must be a reason for the existence of evil. The reason given is freedom.

God has provided humans will free will which is the ability to choose what is moral. All immorality is born from the human capacity to choose evil over good. If the fabric of time does not allow for morality in the Kantian tradition, then it may be that science will eventually challenge the presumption of the accuracy of this philosophical argument for the existence of God.

Does the birth of our universe via time travel preclude the existence of God? According to the philosopher St. Thomas Aquinas, one of the logical means of reasoning which leads to indisputable proof of God is the concept of a "first mover".

Everything in motion was put into motion by something. (For example, something throws a bowling ball down an alley. It does not roll of its own volition.) The Universe is in motion; therefore, the Universe must have a first mover. God requires nothing before; ergo God is the first mover.

However, according to Dr. Gott's the Universe as its Own Mother theory, the Universe could have created itself. If the Universe created itself, then God need not be the first mover. If God is not the first mover, then this argument is not valid.

Despite the philosophical aspects implicit in the dynamics of time, there are also other issues to consider. For the remainder of this chapter, we will assume time to be a real and flowing entity. The following will concentrate on the ethical aspects of the manipulation of time via time travel.

There are five categories of temporal movement that are relevant to those interested in the ethics of time travel: travel forwards and backwards along an original timeline, where each of these is classified by use/non-use of wormholes, and into alternate universes[11]. Each of these can be ethically examined by asking the following questions. Does this type of time travel inflict a willful wrong upon another? Does it take away another's free will? Can we obtain permission to enter the new time? Is this type of time travel more harmful than any other form of adventure or travel? If the answer to the questions pertinent to that particular form of time travel is no, then this thesis argues that that particular form of time travel is ethically sound.

Moving forward along an original timeline without the aid of a wormhole is the first most logical temporal dance to study because it is already being accomplished today. Any time an adventurer travels by plane, train, automobile, or spacecraft, relativity steps in and gives the adventurer a fraction of a second added to their life over that of a stationary

---

[11] Technically, this could be described as going forward, backward, or lateral on another timeline.

observer.

This introduces us to the twin paradox. So-called because it presents the concept of a pair of twins who venture on an unusual journey. This pair of twins, Dannelle and Danny, are 29 years old when friendly aliens kidnap Dannelle. These aliens take her at nearly the speed of light to their home planet about ten light-years away. She stays there for one year and returns.

When she returns, she is several years older than she was at the time of her kidnapping. Her twin sister, however, has aged significantly more than that.

All observers, including both twins, agree that there is now a significant difference in age between the twins, even though they were both born on the same day. This is the twin paradox.

Under normal circumstances, time travel in this capacity is as minute as to be absolutely inconsequential. There are no moral issues with this type of time travel at this level. One might as well talk of the moral issues associated with vehicular travel.

Even as one progresses from minute to moderate to massive, movement along the timeline in the fashion described in the twin paradox offers little in the way of moral obligation, yet there are always issues to watch for. For example, if one were to use a Jupiter time machine to march four hundred years into the future, there is the possibility of throwing off the balance of the system the traveler arrives at. They might bring a disease that had not been seen in that civilization in several hundred years for which there is no longer a cure. Perhaps they had thoughtlessly brought an ecological disease into the ecology that might destroy a delicate balance created in the years since their cocoon in the Jupiter time machine.

It is because of these possible disasters that Dr. John Reuscher, Professor of Philosophy at Georgetown University, recommends that any traveler to the future should attempt to request permission before arriving*. Much like a ship requests permission before docking. If permission is granted, or unattainable, then, according to Dr. Reuscher, this type of time travel is no more ethically questionable than any other means of travel.

Relativity theory also provides means for traveling backward along a timeline without the use of a wormhole. This is accomplished by exceeding light speed or by manipulating space-time in such a way as to trick time into allowing a traveler to outmaneuver light to a destination by means other than speed. The former of these concepts is more morally detrimental than the latter.

---

* Perhaps sending forward a paper request could do this, assuming that a positive answer could be sent back in time.

Speeding into the past has a romantic ring to it, but potentially dire consequences. There are several different ethical issues of concern here. Each of the following issues is relevant to nearly all means of time travel but shall be discussed under the heading of the most relevant method of time travel.

The previous chapter discussed the Butterfly Effect, which is an expression of chaos theory, so it only requires a brief mention here. The theory explains changes made to initial conditions, for example, that happened on the timeline between "now" and the past, could dramatically affect future events, therefore, like a gentle touch on an intricate spider's web, changes may be made that will have a far-reaching impact: good, bad, and indifferent. Regardless, as these changes are universal, wielding the power of time travel means wielding power to alter innumerable others' past, and therefore their futures.

The Butterfly Effect is one of the more difficult effects of time travel to ethically evaluate. It does not preclude acts of willful wrong against another, nor the revocation of another's free will. Yet it does not prevent it. Permission is unfeasible as there is no means to get permission from all those affected. However, one can argue that the Butterfly Effect produced from time travel is no different from the Butterfly Effect formed from any other form of travel.

Consequently, it is not the Butterfly Effect that warrants evaluation, but the intentions of the travelers and the method they choose to use that need ethical evaluation.

The popular name for chronology protection is the grandfather paradox, a popularized concept explored in many books and movies. The plotline will invariably follow a set generic theme. The hero will go back in time, probably to try to set right some past wrong. He will venture into the time when his grandfather would be the same age as the time-traveler and the time-traveler will therefore not recognize him. After several plot twists, the hero will find himself unknowingly about to create the demise of his grandfather. Fortunately, Mother Nature herself, or Fate, will step in and keep the hero from killing his grandfather. This, in turn, prevents the grandfather paradox that states that the death of the grandfather will prevent the birth of the father, preventing the birth of the hero that in turn will prevent the time travel resulting in the death of the grandfather.

The Grandfather paradox allows for the intent of a willful wrong, but not the implementation. The same can be inferred about the alteration of another's free will. These are the questions most relevant to the grandfather paradox and these answers illustrate that there is no serious ethical issue associated with this paradox.

The information protection paradox (Davies 2001, 102) explains the possibility of information or knowledge created from nothing. For example, a time traveler could go into the past and provide a scientist with the information needed to create a time machine. That scientist would then use that knowledge to create a time machine. That time machine then turns out to be the same as the one the original time traveler uses to go back in time to provide the information that ultimately created the vehicle used to travel back in time. In essence, knowledge emerges from nothing.

From knowledge to wealth, the wealth protection paradox is quite like the knowledge protection paradox. In this case, it is physical matter which emerges from nothing. For example: A woman takes a bar of gold that her parents gave her and travels back in time to her parents' wedding day. If she leaves it on her parent's porch on the day of their wedding with a note saying, "a wedding gift", her parents would be in proud possession of a bar of gold with a circular lifeline. Any known machinery or mechanism did not create it; in fact, it was never created.

A bar of gold or a piece of information with a circular lifeline poses no ethical dilemmas. However, according to Dr. John Reuscher, if you were to replace that bar of gold with a small child, there would instantly be a problem. While the information

protection paradox cannot harm inanimate objects
by losing their future, it can harm a sentient being. In
this type of situation, a child (or any living creature)
is deprived of their freedom by the loss of their
future ability to make choices. From the Kantian
perspective, the deprivation of freedom is one of the
most egregious harms that may be inflicted upon a
person. Therefore, this automatically falls under the
question of the revocation of another's free will.

If one controlled the power of time, might one allow
power to overwhelm one's sense of morality? Extreme
power corrupts, and power over time and causality
could take mortality away from the hands of the grim
reaper and into the hands of the powerful.

Playing God in this way can be exemplified by the
following example. A woman and her child are struck
and killed by a drunk driver on their way to soccer
practice. Ten years later her husband is introduced to
a keeper of a time machine. The husband offers the
keeper a large sum of money to allow him to go back
ten years and adjust the past so that his wife and
child take an alternate route on the night in question.

The implications of this type of alteration to the past
are staggering. This poses an example of how the rich
and powerful can use their influence to continue
to improve their situation while leaving the poor
in a stagnating or worsening condition because an

indigent person could never bribe a "keeper" to allow them access to the past.

Imagine replacing the husband in the story with an entire government. Now replace the lost wife and child with a battle and the tide of the entire world order can be shifted in just one moment or replace both of the above with a solipsistic personality. This could lead to endless cycles of travel to the past to repair any/many previous perceived wrongs or errors.

Now assume that it is not just the rich and powerful that have control over time. Imagine that there is free and easy access to time machines for anyone who would choose to explore or change the past. Populations would explode, as accidental deaths become a phenomenon of the past. The past would no longer be sacrosanct as old errors are righted, and these righted errors are extrapolated into new problems. The "now" would become just a fleeting blur between possible futures.

Both pay-per-access and free and easy access to time machines pose several serious ethical problems. Either type of access allows for the manipulation of the past or future in such a way that a willful wrong or the annihilation of another's free will becomes a likelihood. Additionally, these methods of time travel scoff at the idea of asking permission to visit another time because it would be logistically difficult to ask permission for large numbers of time travelers.

The final paradox associated with methods of time travels that utilize relativity theory could be called the observation effect. At the heart of quantum theory is the notion that observation affects the observed. The tale of Schrödinger's cat demonstrates this. On the macroscopic scale, it is important to question whether a similar effect might take place.

In experiments, scientists have recorded changes that are brought about by the mere act of observation. This is currently found at the microscopic level - but experiments have never been done to show that an observer from the future has no impact on the observed in the past. If then, the observer does have an impact, it may be that the observer affects the free will (or at least alters the assumption of free will/right to choose) of the observed by the mere act of observation. However, if it is proven that because the observer is there (in the past), then the observer must have always been there (in the past), then there is no impact on free will because that which originally happened (in the past) will become what always happened (in the past), regardless of the observer.

If a scholar were to transport herself back in time to observe a great event in history, is it possible for her to have no impact on the place around her? Unless she was observing from within another dimension

(unlikely), then there is no way for her to keep the
environment sterile from her presence. Imagine
she takes one step and kills a resting butterfly. The
Butterfly Effect is a powerful force of nature.

Accelerated forward movement along a timeline
produces no significant implications short term but
has catastrophic potential long term. The most likely
practical use of time travel to the future includes the
use of rocket ships and Jupiter time machines. Use of
a rocket ship with near light-speed velocity, or travel
via a Jupiter time machine has the potential to take
someone hundreds (or more) of years into the future.
The impact of the sudden arrival of a time traveler
into a new environment could induce dramatic
ramifications. It might be something like the impact
of the American colonialists on the Native American
Indians.

Disease, which had been wiped out centuries ago,
could re-infect and spread quickly and brutally.
Or, perhaps, in this future, entire concepts such as
competition and war had been eliminated. A time
traveler, possibly through ignorance, and probably in
all innocence, could reintroduce these concepts and
wreak havoc with a civilization.

Another possible scenario has the future civilization
in such a tightly balanced economy that the
introduction of even one person could trigger a

cataclysmic chain reaction devastating the world
(galactic) economy. A similar threat could be made to
the ecology of the future system. The introduction,
for example, of a pet dog or a stow-away rat could
destroy a perfectly balanced ecosystem.

All of these reasons may lead to the presumption
that traveling to the far future using a rocket ship
or a Jupiter time machine is unethical. Not so,
argues Dr. Reuscher. According to Dr. Reuscher, the
implications of this type of time travel are no more
morally questionable than those faced by an explorer
traveling to unknown lands. However, he qualifies
this answer by saying that one must aim to protect
whatever future one arrives at. If it is possible to send
a message to the future asking permission to visit,
then that must be done to maintain the most ethical
approach to time travel.

In addition to traveling forward along an original
timeline utilizing constructs of relativity theory,
there is the possibility to travel into the future by
manipulating space-time in a quantum way. Such
travel would be accomplished using wormhole
technology. The advantage to this type of time travel
is that the same method can be used to go into
the future as well as into the past. It also allows for
minimal space travel and uses very little of one's
personal time. This means that one could go a million
years into the future without aging at all.

Travel to the future through a wormhole would produce the same paradoxes and ethical issues as time travel via a rocket ship or a Jupiter time machine but would also produce other conundrums. Wormhole travel offers a few options that other time travel does not. One of these is the ability to travel back and forth along a single timeline without the inconvenience of taking a long time to do so. The other is that it grants the ability to jump over time, that is, wormholes provide a way to get to one specific time without having to pass through all of the time in-between. Finally, wormholes offer the advantage of accuracy. One mouth of the wormhole is anchored to a specific place and time so any traveler will have an accurate idea of where they will end up - at least in one direction.

Although wormhole travel offers advantages over other types of time travel, it does open the door to ever more troubling corruption issues. Wormholes allow time travelers to jump from one time to another. Unlike other forms of time travel, however, wormhole travel provides a means to jump over known points in time. This means that if a person or ethnic/religious/political group had access to wormhole travel while another did not, then the composition of the world could be changed. If, for example, a wormhole time machine had been erected before World War II, then a future generation could

go back in time and transport all the inhabitants
of the concentration camps and relocate them to a
better time.

What would this mean to the world? The implications
are far-reaching. That war would have claimed far
fewer lives. However, the cost of those saved lives
would manifest itself in the loss of lessons learned
from war crimes. History is bound to repeat itself if
no one listens to its lessons.

Additionally, that generation's impact will be gone
from today's world. The offspring of that generation
will also be misplaced in time. Good deeds as well
as bad, follow the Butterfly Effect; there is no way to
evaluate the impact of these missing lives.

Dimensional travel must be included in any informed
discussion of time travel because dimensional travel
may be time travel. One way to envision this is
through quantum theory. Quantum theory postulates
that a new dimension might be created for every
action taken, then it is reasonable to assume that the
time and place that you want to go to exist in one or
more alternate dimensions.

Another facet to this concept may be drawn from
chaos theory. Chaos theory proclaims that every
motion reverberates far and wide. This can be
translated into the idea that time travel may create

alternate universes.

This interpretation rests on the following example. A time traveler in 2022 enters a wormhole time machine and exits in 1900. Upon his exit, a mosquito lands on his arm and he kills it before it has a chance to bite him.

In the original timeline that the time traveler came from, that mosquito had West Nile virus and had led to the death of a prominent businessman. That businessman had been one of the financial backers of wormhole research. His death postponed the creation of a useable wormhole by nearly twenty years. Assuming a time loop is averted, the logical conclusion is that the timeline splits or a new universe is created to solve the implicit paradox.

This incident has created one of two possible scenarios. The first is that there has been created a break in the natural timeline whereby the alternate timeline carries on with the business mogul who dies prematurely, and the time traveler is stuck in the "past". The second is that the ripple effect caused by the damage to the past creates an alternate universe a la quantum theory.

Either scenario alters the lives of every sentient being in the universe. It could be that this is an immoral act in that the alteration of another being's life without their consent may be deemed an inherent wrong. However, with help from the Butterfly Effect, it is easy to see that any action alters the life of other beings without their consent. As you picked up the hardcopy of this book to read it, the change in air movement generated by the flap of the papers has started on its path to change the future of the world. It may be argued that the universe hopping form of time travel is not more ethically challenged than any other form of transportation.

The other way to enter an alternate universe would be through traversing a black hole. The difference here is that the traveler's actions are not creating the alternate universe, the traveler's actions are only breaching the divide between concurrently existing, but separate universes.

Entering an alternate universe of this type opens questions that all other types of time travel can avoid. Does this universe operate under the same laws of physics? If so, is the law of the conservation of energy being upset by the addition of a previously nonexistent being? Do humans exist/can humans exist in the environment? Is it ethical to traverse a black hole into another universe?

It is currently impossible to answer any of those questions, except the last one. Is it ethical to traverse a black hole into another universe? For the answer to this question, it seems we should rely on the criteria used for all other types of time travel.

• Is this travel willfully doing wrong to another? No.

• Is this travel taking away another's free will? No.

• Can permission be obtained before attempting this type of travel? No.

• Using reductionism, can the impact of this type of time travel be compared to another type of travel (for example-driving a car)? Yes.

The answers to these questions tell us that traversing a black hole into another universe is ethically acceptable.

Epilogue/Conclusion

Now is the time to re-evaluate what should be considered impossible. Science has breathed life into yesterday's science fiction; men do walk on the moon and fly without wings. Every day something that was impossible becomes that which is.

Relatively speaking, time is no longer a universal constant. It is greatly affected by both speed and gravity. Time's clock is individually set. These concepts are the foundation for several forms of time travel.

Quantum theory holds even more surprises. Black

holes, wormholes, and alternate universes exist. Not only do they exist, but they offer additional means of time travel.

Chaos reigns in our reality. Chaos theory allows for both regularity from complexity and chaos from simplicity while always being based on sensitivity to initial conditions. This then creates the Butterfly Effect that is a scientifically proven effect that bears heavily on the ethical implications of time travel.

 Time travel may produce paradoxes, time loops, split timelines, and alternate universes. Can it be ethical? The answer is yes. Ultimately, however, each circumstance must be individually examined. Actions can seldom if ever, be ethical if willful wrongs are committed against others or if free will is limited. This book has shown that most forms of time travel are no more detrimental than other methods of travel. Therefore, what is the future of the impossible? Time travel.

Bibliography

Print Sources:

Barrow, John D. Impossibility – The Limits of
Science and the Science of Limits. Oxford: Oxford
University Press, 1998. Pages 12, 195-196.

Barrow, John D. Theories of Everything – The Quest
for the Ultimate Explanation. Oxford: Clarendon
Press, 1991. Pages 39-41, 44-60, 62-64, 149-150.

Capra, Fritjof. The Tao of Physics. New York: Bantam
Books, 1984. Pages 147-174.

Coveney, Peter and Roger Highfield. The Arrow of
Time – A Voyage Through Science to Solve Time's
Greatest Mystery. New York: Fawcett Columbine,
1990.

Davies, Paul. About Time – Einstein's Unfinished
Revolution. New York: Simon and Schuster, 1995.
Chapters 10 and 11.

Einstein, Albert. Relativity. New York: Crown
Publishers, Inc (no printing date given).

Gardener, Robert. Experimenting with Time. New
York: A Venture Book,

Gleick, James. Chaos Making a New Science. New
York: Viking, 1987.

Greene, Brian. The Fabric of the Cosmos: Space,
Time, and the Texture of Reality. New York:
Random House, Inc., 2004.

Gribbin, John. Unveiling the Edge of Time. New York:
Crown Trade Paperbacks, 1992. Pages 28-57 and
115-213.

Hall, Nina, ed. Exploring Chaos – A Guide to the New
Science of Disorder. New York: WW Norton and
Company, 1994.

Hawking, Stephen. The Universe in a Nutshell. New York: Random House, Inc, 2001.

Hawking, Stephen and Roger Penros. The Nature of Space and Time. Princeton: Princeton University Press, 1996. Pages. 27-31.

Kant, Immanuel. Prolegomena to any Future Metaphysics. Indianapolis: The Bobbs-Merrill Company, Inc., 1950. Pages 13-27.

Kant, Immanuel. The Philosophy of Kant. New York: Random House, Inc., 1977. Pages 24-29 and 140-264.

Macvey, John. Time Travel – A Guide to Journeys in the 4th Dimension. Chelsea: Scarborough House Publishers, 1991.

Overbye, Dennis. From Space, A New View Of Doomsday: How Mysterious Dark Energy Might Blast the Universe Apart. The New York Times, Tuesday, February 17, 2004, D1-2.

Peterfreund, Denise. Great Traditions in Ethics, 7th Edition. Belmont: Wadsworth Publishing, 1992.

Pickover, Clifford A. Chaos in Wonderland – Visual Adventures in a Fractal World. New York: St. Martin's Press, 1994.

Prigogine, Ilya. The End of Certainty – Time, Chaos, and the New Laws of Nature. New York York: The Free Press, 1997.

Thorne, Kip. Black Holes and Time Warps – Einstein's Outrageous Legacy. New York: WW Norton and Company, 1994.

Kip Thorne and Ulvi Yurtsever, "Wormholes, Time Machines, and the Weak Energy Condition," Physical Review Letters 61, no. 13 (September 1988): 1446-1449.

Zimmerman, Barry E. and David J. Zimmerman. Why Nothing Can Travel Faster than the Speed of Light and Other Explorations in Nature's Curiosity Shop. Chicago: Contemporary Books, 1993. Pages 79-84 and 143-154.

Internet Sources:

Spacetime Wrinkles Map, [resource on-line]; available from http://archive.ncsa.uiuc.edu/Cyberia/Expo/ numrel_nav.html; Internet; accessed January 30, 2004.

A Panorama of Fractals and Their Uses, [resource on-line]; available from http://classes.yale.edu/

fractals/Panorama/Nature/Coastlines/PolyTest.html;
Internet; accessed February 16, 2004

The Story of Chaos Theory, [rsource online]; available
from http://members.fortunecity.com/kokhuitan/
chaos.html; Internet; accessed January 29, 2004

Time Travel and Paradox, [resource on-line]; available
from http://homepage.mac.com/billtomlinson/
newtt.html; Internet; accessed March 18, 2004

Deterministic Chaos, [resource on-line]; available
from http://pespmc1.vub.ac.be/CHAOS.html;
Internet; accessed January 29, 2004 This is a
website on deterministic chaos.

Author is F. Heylighen. Created on October 14, 1998
and modified on April 26, 2002.

Brian Greene, [resource on-line]; available from http://
phys.columbia.edu/faculty/greene.htm; Internet;
accessed March 4, 2004

Breaking the Light Speed Limit, [resource on-line];
available from http://science.howstuffworks.com/
news-item6.htm; Internet; accessed March 1, 2004

Brian Greene, Author of the Elegant Universe,
[interview on-line]; available from http://
superstringtheory.com/people/bgreene.html;

Internet; accessed March 4, 2004 This is a branch of
the Official String Theory web site.

Ockham's Razor, [resource on-line]; available from
http://whatis.techtarget.com/definition/0,,sid9_
gci212684,00.html; Internet; accessed March 18,
2004

Spirofractal Gallery, "Fractal Tour", fern.jpg, [resource
on-line], available from http://www.alunw.freeuk.
com/fractaltour.html; Internet; accessed February
13, 2004

Biographies: Immanuel Kant., [resource on-line],
available from http://www.blupete.com/Literature/
Biographies/Philosophy/Kant.htm; Internet;
accessed March 31, 2004

http://www.ccs.uky.edu/~dejan/work/news/Gott_
interview/J.Richard.Gott.III.eng.html March 4, 2004

http://www.ccs.uky.edu/~dejan/work/news/Gott_
interview/universe_tree.jpg March 4, 2004 self
creating universe picture

Proofs for and Against the Existence of God,
[resource on-line], available from http://www.cftech.
com/BrainBank/OTHERREFERENCE/PHILOSOPHY/
ArguAbtGod.html; Internet; accessed March 31,

2004

Light Speed, [resource on-line], available from http://
www.electrogravityphysics.com/html/speed_of_
light.html; Internet; accessed March 1, 2004

Author Jonathan Leake. Created June 4, 2000.

http://www.friesian.com/paradox.htm March 18,
2004Time Travel Paradoxes. Summary of time
travel paradoxes.

Time's Dependence on Space: Kant's Statements and
Their Misconstrual by Heidegger, [resource on-line]
available from http://www.focusing.org/gendlin_
times_dependence_on_space.html; Internet;
accessed March 19, 2004

Author is Eugene T. Gendlin.

Deterministic Chaos, Fractals, and Quantumlike
Mechanics in Atmospheric Flows, [resource on-
line], available from http://www.geocities.com/
CapeCanaveral/Lab/5833/canjphy/canj.html;
Internet; accessed January 29, 2004

Author is A. Mary Selvam.

Concise Encyclopedia of Mathematics, "Sierpinski
Sieve", sierpinski_clear.gif, [resource on-line],

available from, http://www.itu.dk/bibliotek/
encyclopedia/math/s/s283.htm; Internet; accessed
February 13, 2004

Superstring Theory, [resource on-line], available from
http://www.lassp.cornell.edu/GraduateAdmissions/
greene/greene.html; Internet; accessed March 4,
2004

Chaos Theory, [resource on-line], available from http://
www.ncst.ernet.in/kbcs/vivek/issues/11.4/puneet/
puneet.html; Internet; accessed January 29, 2004

Author is Puneet Srivastava.

Physicists Slow Speed of Light, [resource on-line],
available from http://www.news.harvard.edu/
gazette/1999/02.18/light.html; Internet; accessed
March 1, 2004

Author William Cromey.

The Elegant Universe, [video on-line], available at
http://www.pbs.org/wgbh/nova/elegant/program.
html; Internet; accessed January 30, 2004

Brian Greene.

Time Travel, [resource on-line], available at http://
www.pbs.org/wgbh/nova/time/; Internet; accessed

January 30, 2004

PBS website.

Signals Exceed Light's Speed, [resource on-line],
    available at http://www.photonics.com/spectra/tech/
    XQ/ASP/techid.879/QX/read.htm; Internet; accessed
    March 1, 2004

Tachyons, [resource on-line], avaiable at http://www.
    physics.gmu.edu/department/research/tachyons.
    htm; Internet; accessed January 30, 2004

Noisy Chaos, [resoucre on-line], available at http://
    www.santafe.edu/projects/CompMech/papers/
    NCTitlePage.html; Internet; accessed January 29,
    2004

James P. Crutchfield.

http://www.shef.ac.uk/~phil/courses/207/10-Morality.
    pdf March 31, 2004

Kant on Space and Time, [resource on-line], available
    at http://www.soci.niu.edu/~phildept/Sytsma/
    KANTS&T.html; Internet; accessed March 19, 2004

Why Did Strings Enter the Story? [resource on-line],
    available at http://www.superstringtheory.com/
    basics/basic3.html; Internet; accessed January 30,

2004

Choas, [resource on-line], available at http://www.
    zeuscat.com/andrew/chaos/chaos.html;available
    January 29, 2004

Interviews conducted by the author:

Reuscher, John, interview by author, Director of
    Graduate Studies, Philosophy Department,
    Georgetown University. 218 New North, Box
    571133.

About the Author

The Future of the Impossible, the Physics and Ethics of Time Travel is based on Dannelle Shugart's thesis for the Georgetown University's Master of Arts in Liberal Studies program. She wrote it in part to give people who are interested in physics and time travel, but who are not scientists, a way to learn about and enjoy the topic.

Ms. Shugart is not a trained scientist but believes that much of the world is interconnected and that to learn about science is to understand art better. Just as learning about mathematics allows a deeper connection to music and so on. The more diversly you learn, the more fascinating and evident disparate connections become.

In addition to her interest in time travel, the author enjoys travel, cozy mysteries, parenting, entrepreneurship, photography, and lifelong learning. She also believes in the value of being "weird".

She really hopes you enjoy this book.